IN THE
NATIONAL INTEREST

General Sir John Monash once exhorted a graduating class to 'equip yourself for life, not solely for your own benefit but for the benefit of the whole community'. At the university established in his name, we repeat this statement to our own graduating classes, to acknowledge how important it is that common or public good flows from education.

Universities spread and build on the knowledge they acquire through scholarship in many ways, well beyond the transmission of this learning through their education. It is a necessary part of a university's role to debate its findings, not only with other researchers and scholars, but also with the broader community in which it resides.

Publishing for the benefit of society is an important part of a university's commitment to free intellectual inquiry. A university provides civil space for such inquiry by its scholars, as well as for investigations by public intellectuals and expert practitioners.

This series, In the National Interest, embodies Monash University's mission to extend knowledge and encourage informed debate about matters of great significance to Australia's future.

Professor Margaret Gardner AC
President and Vice-Chancellor,
Monash University

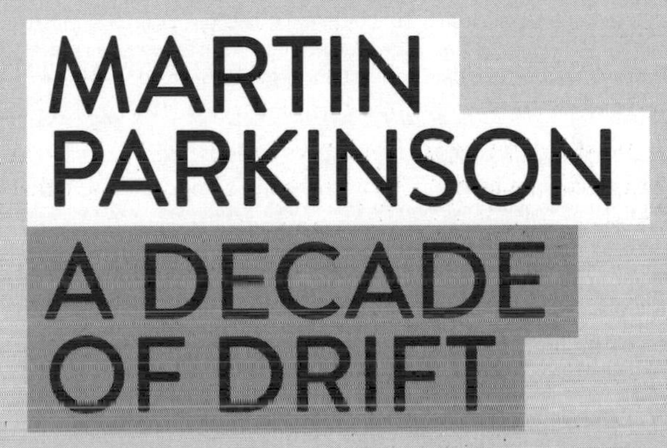
MARTIN
PARKINSON
A DECADE
OF DRIFT

MONASH
UNIVERSITY
PUBLISHING

A Decade of Drift

Monash University Publishing
Matheson Library Annexe
40 Exhibition Walk
Monash University
Clayton, Victoria 3800, Australia
https://publishing.monash.edu

Monash University Publishing brings to the world publications which advance the best traditions of humane and enlightened thought.

ISBN: 9781922464095 (paperback)
ISBN: 9781922464118 (ebook)

Series: In the National Interest
Editor: Louise Adler
Project manager & copyeditor: Paul Smitz
Designer: Peter Long
Typesetter: Cannon Typesetting
Proofreader: Gillian Armitage
Printed in Australia by Ligare Book Printers

A catalogue record for this book is available from the National Library of Australia.

The paper this book is printed on is in accordance with the standards of the Forest Stewardship Council®. The FSC® promotes environmentally responsible, socially beneficial and economically viable management of the world's forests.

A DECADE OF DRIFT

The history of economic reform in Australia over the last decade and a half stands in sharp contrast to the two decades before. The early 1980s to the early 2000s was a time of great political stability, with bipartisan support for the general direction of policy and broad community acceptance of the need for reform, if not its detail and pace. The actual reforms, pursued under governments led by both major parties, set the scene for a record twenty-eight years of uninterrupted economic growth. This is not to suggest that reform was easy—it wasn't—or that success was preordained. But the political centre and the economics profession were aligned on the direction of reform.

Paul Kelly has described three unique factors as giving rise to the post-1983 reform agenda:

- widespread recognition that the status quo had failed, a sentiment driven by the deep recessions of the 1970s and early 1980s
- a consensus in the political system in favour of new ideas such as the utility of markets and the need to deregulate, wind back protection, increase savings and impose more discipline on the public sector
- a political culture that was able to prioritise the national interest.[1]

While correlation does not imply causation, it is hard to conceive of that reform agenda being delivered in an environment of dramatic political instability. One can imagine, though, that successful reform could contribute to political stability, illustrating both government's intent to improve the lot of citizens and its competence to deliver.

Whatever the relationship, it has been clear for some time that political stability, and Kelly's three factors, have been absent from the Australian policy landscape.[2]

The change in political stability is illustrated by my own experience. Having joined Treasury as a cadet in 1980, my working life for twenty-eight years saw Australia served by just four prime ministers: Malcolm Fraser, Bob Hawke, Paul Keating and John Howard. In contrast, the last twelve years of my professional public service career saw a giddying six occupants of the Prime Minister's Office: Kevin Rudd, Julia Gillard, Kevin Rudd again, Tony Abbott, Malcolm Turnbull and Scott Morrison.

While each prime minister has had a view of the national interest at heart, the seventeen years since John Howard's last election victory have been characterised by an absence of a vision for Australia that the public has found compelling. Australia's voters have neither been captured by the need to respond to the circumstances we confront, nor excited by a future unfolding. Our social cohesion has been damaged and the gaps in our society wedged further apart by a fracturing of the political centre.

It is too convenient to blame these circumstances on the supposed inadequacies of Australia's political and bureaucratic class, whether federal, state or local. The sharp decline in trust in many of our institutions of power and influence—be they government, business, non-government organisations (NGOs; for example, churches), or the traditional media—suggest something far deeper is at play, not only in Australia but across the Western world. While the causes are doubtless complex, it is highly likely that social media, and the ability to source information from an echo chamber of people sharing one's own views, has contributed to this situation. Whatever the reasons, a lack of trust has become a significant impediment to good decision-making.

Indeed, the Edelman Trust Barometer has in recent years suggested Australians have had consistently low levels of trust in government and politicians, with the gap in trust between what it refers to as 'the informed public' and 'the mass population' now at record levels. Importantly, no Australian institution—government, business, media or NGOs—was seen as both ethical and competent when the most recent survey was conducted towards the end of 2019.

In normal circumstances, such disillusionment would be a cause for great concern. But as Australia attempts to start the recovery from the COVID-19–induced economic shock, the current circumstances risk hobbling our ability to take the actions necessary to return to a pathway of growth in economic activity, employment, incomes and living standards.

Moreover, the standing of the economics profession has not recovered from the shock of the 2008 global financial crisis (GFC). There has been no broad consensus on the needed direction of policy, and much of the deregulatory, market-oriented policy direction—in particular, the free movement of capital, people, goods and services—that underpinned Australia's unprecedented twenty-eight years of uninterrupted economic growth, is now politically contested. In Australia, a consensus on the policies required to abate greenhouse gas emissions has also remained out of reach. The economics profession, with a few notable exceptions, has been relatively ineffectual in its policy contribution during the climate change wars of the last decade.

The failure to return to a strong and sustainable growth pathway risks consigning Australians to declining relative living standards and levels of government debt that sharply curtail policy options and impose a high burden on future generations. It also means that governments may have less capacity to manage future economic shocks, that Australia may be unable to grasp the opportunities provided by living in the most dynamic and exciting part of the world, and that we are unprepared for the strategic, economic and other challenges we will confront in the Indo-Pacific.

This book looks back over the last decade or so to better understand a key, perhaps *the* key, public policy failure of modern times: Australia's inability to settle on a clear policy approach to the challenge of climate change. Perhaps no other single issue so clearly illustrates the absence of Kelly's three factors. Arguably, no other single policy failure has contributed more to public disillusionment about the competence and beneficent intent of government.

A decade of drift, of climate policies proposed and abandoned, of science acknowledged and then ignored, means Australians are more vulnerable to social and economic dislocation and disruption caused by rising temperatures. It has meant higher electricity prices for households and business, and lost investment opportunities and jobs, and has made the eventual adjustment process more expensive. It has also left Australia with even more assets—roads, bridges, houses, factories, livestock—at risk of extreme climate events, or in danger of being stranded and abandoned before the end of their physical lives. The risks to human health, indeed to Australians' lives, can no longer be dismissed as part of the natural cycle of weather.

This is not to suggest Australia has achieved nothing in terms of broader policy over the last decade or so. To the contrary, across economic, social and strategic domains, progress has been made.

The Rudd government helped establish the G20 as a leaders' forum, drive a globally coordinated response to the GFC, and avoid a GFC-induced recession. The Gillard government introduced the National Disability Insurance Scheme and developed a revised strategy for engaging with Asia. The Abbott government started the process of budget repair and negotiated free trade agreements with China, Japan and South Korea.

The Turnbull government landed a refugee deal with the Obama administration that led to the closure of Manus Island and falling numbers of detainees on Nauru. It instigated a national innovation strategy, managed the tumultuous early days of the Trump administration, delivered same-sex-marriage equality, established the first cybersecurity strategy and initiated the Pacific Step-up. The Morrison government, which before the COVID-19 crisis was on the verge of delivering the first surplus since 2007–08, managed well the initial medical and economic response. It has also focused further attention on countries closer to home, strengthened the long-overdue

Pacific Step-up and begun charting a more independent foreign policy in the face of US–China strategic competition.

Throughout the period since 2013, under the Abbott, Turnbull and Morrison governments, there has also been a consistent theme of Australia needing to develop a greater sovereign defence capability, an approach which builds on the Rudd government's 2009 Defence White Paper.

What is conspicuous, though, when set against this record of achievement, is the multiple attempts at climate policy. Starting with Kevin Rudd's Carbon Pollution Reduction Scheme (CPRS), defeated in the Senate by one vote, Australia experimented with Julia Gillard's carbon pricing mechanism (CPM), which was in turn abolished and replaced by Tony Abbott's Direct Action.

Malcolm Turnbull made two attempts: an initial Clean Energy Target (CET) developed by Australia's Chief Scientist, Professor Alan Finkel, and later the National Energy Guarantee (NEG). Notwithstanding broad business and community support for the NEG, opposition to it ultimately became a rallying point for removing Turnbull from the Prime Minister's Office.

Both the government and, more recently, the ALP Opposition have now disavowed carbon pricing, turning their back on the cheapest, most efficient way of reducing emissions, in favour of more direct, costly and risky interventions. The toxicity of carbon pricing has restricted the scope for policy development and innovation, with the Morrison government embarking on the development of a technology road map in the hope that a focus on the technologies needed to reduce emissions can provide some common ground for action.

A technology road map, if done well, can provide a useful complement to the pricing of emissions, but it is not a substitute for either the incentive effects or the information transfer provided by a market price. If done badly, it can be costly to the public purse, distort public and private investment, and result in Australia forgoing opportunities for future growth and jobs.

This is not the place to pass judgement on the government's emphasis on a road map—it will take time to assess the impact of what's been proposed. But the fact that Australia's only policy options are now so limited, and such a long way from a comprehensive 'first best' strategy, highlights the challenge of achieving policy consensus in this area.

Whether considering climate change or the immediate need to rebuild our post-COVID-19 economy, Australia needs some degree of consensus on priorities and action. If we have a governing class that, irrespective of where they sit on the political spectrum, pretends that complex issues are simple, rather than acknowledging that some problems require difficult decisions, the Australian public will be short-changed. Excessive effort spent admiring the problem—acknowledging the complexity but failing to act—is equally counterproductive.

Unfortunately, working against the achievement of solutions to either of these two great challenges is the prevalence of 'short-termism' and the 'gotcha' mentality—leading to a fear that trying and failing to fix critical challenges, no matter how difficult it will be to succeed, will lead to accusations of incompetence or weakness. This feeds back into the erosion of public trust in institutions and government in particular.

It is not the object of this book to assess whether the apparent inability, or unwillingness, to be transparent on the magnitude of the challenges, or to invest in building community support for difficult policy decisions, reflects a lack of conviction or courage, or simply political realities. Ironically, though, it may be that the experience of COVID-19—the transparency shown in describing the challenge, the creation of the National Cabinet, the obvious recognition and embrace of expert advice—may help to embed these aspects in our policy debates and turn around this trust deficit. If so, it will be to the benefit of all Australians.

GOOD POLICY = GOOD POLITICS

The capacity to prosecute and advocate for significant policies and reforms in the public sphere has long been seen as a mark of good government and good governance—the mantra that good policy is good politics! It is indisputable, however, that raising the need to address climate change in a realistic, balanced manner leaves politicians as easy targets for their opponents at both ends of the spectrum—those accusing them of doing too much or too little. Only some form of consensus around the political centre can deliver Australia the policy responses that it needs to sustain over decades to manage the transition of our economy to a low-emissions world. The consequences of failure are almost too hard to conceive.

In a low emissions world, one to which Australia had not adapted, our economy would be disproportionately dependent on increasingly uncompetitive industries, with investment, jobs and living standards all in relative decline. This would affect all Australians, but it would have acute impacts on certain geographic areas where our policy short-sightedness had prevented us managing the transition when we had the opportunity. We would have reduced the capacity to generate high-quality, high-wage jobs and opportunities for our citizens. Other countries could penalise our emissions-intensive exports, and our ability to influence international decisions in our national interest would be eroded.

The alternative—commencing a measured, sustained transition to a low-emissions world—will not deliver nirvana, but it offers prospects and opportunities for all and, importantly, allows for a just transition in the industries and regions most affected. This need for consensus, and recognition of the consequences of failure, also apply to the achievement of a post-COVID-19 recovery strategy.

Economists are not immune from this lack of consensus and confidence. Should economic policymakers be prioritising the regulation of 'big tech' or should they be trying to improve the lives of Indigenous Australians? Is the state of our education or industrial sectors most critical, or should climate change be at the forefront of our thinking? The issues that can be juxtaposed are almost infinite, reflecting the individual priorities of those putting forward the ideas. While every one of these issues may be important, they can lead in very different directions. The challenge is to develop policy frameworks and political narratives that are internally coherent and can be communicated to reasonable people, thereby enabling the tackling of multiple issues simultaneously.

Ironically, the need for a post-COVID-19 recovery strategy may well offer the opportunity for Australia to accelerate, and better manage, the transition to a low-carbon economy. This transition is underway, albeit somewhat haphazardly, driven by citizens, investors and business. Again, though, much of the momentum is coming from overseas—a reflection of our repeated policy failures—and the risk is that Australia is a laggard rather than a leader in grasping the new opportunities that are emerging.

There is no magic formula for policy success, but in what follows, readers will hopefully discern some of the key elements for success and recognise some common but complex causes of failure. Only by drawing on these lessons can we hope to achieve the consensus needed to return Australia to a post-COVID-19 growth path, one that is both sustained and environmentally sustainable, and provides citizens with the opportunity to lead lives of value.

THE CLIMATE WARS

The history of economic policymaking in Australia has been one of governments progressively removing themselves from allocative decision-making in favour of market-based responses. By the early 2000s, many politicians, policy advisers and commentators saw a move to pricing greenhouse gas emissions as a logical progression of Australia's approach to reform.

Prior to the Gillard government's legislation of the carbon pricing mechanism, which came into operation on 1 July 2012, Australia had never implemented a national market-based reform to manage a global externality, such as that resulting from greenhouse gas emissions.[3] However, debate over the use of an economic instrument to manage Australia's emissions had, by that time, already had a long history.

In 1995, John Faulkner, the then minister for the environment in the Keating Labor government, proposed that the government consider a low-level carbon levy, set at $1.25 a tonne of carbon dioxide–equivalent greenhouse gases. However, facing strong opposition, it was ultimately put aside, with the government instead introducing the Greenhouse Challenge scheme with voluntary participation by industry.

After negotiating the Kyoto Protocol in 1998,[4] Robert Hill, the then environment minister in the Howard Coalition government, commissioned the Australian Greenhouse Office to develop four discussion papers on design aspects of a possible emissions trading scheme (ETS)—these were released in 1999.[5] This was the first detailed exposition in Australia of what such a scheme might entail. While there was an expectation that these papers would be followed by further policy work, leading to the implementation of an ETS, the government decided in 2000 not to proceed with further development of that policy.

In 2002, the Australian Greenhouse Office released a final paper on emissions trading to inform the Council of Australian Governments Energy Market Review.[6] That review, chaired by Warwick Parer, a former minister for energy and resources in the Howard government, recommended the introduction of a national ETS. In 2003, an ETS proposal was again taken to the Howard Cabinet, with the support of four ministers, including then-treasurer Peter Costello, but again the decision taken was not to proceed. Partly as a result, in 2004 the eight state and territory governments jointly established a taskforce to develop a national ETS design. The National Emissions Trading Taskforce released two discussion papers and maintained an ongoing dialogue with stakeholders over possible scheme design from 2005 to 2007, before providing a final report to the Garnaut Review in December 2007.

In 2006, the Howard government established the Task Group on Emissions Trading, chaired by Dr Peter Shergold, the secretary of the Department of the Prime Minister and Cabinet, to reconsider the possibility of an ETS for Australia. I led the secretariat that supported the task group.

The initial reaction of the media and environmental groups to the establishment of the task group was one of scepticism. Many saw it as a smokescreen to allow the government to continue to avoid action, pointing in particular to the fact that a majority of the members were from the business community.[7] What critics didn't know was that, at its very first meeting, every member indicated a strong presumption in favour of acting to reduce Australia's emissions and a desire not to be bound by the terms of reference provided by the government.

The Shergold Report was released in May 2007 and, to the surprise of many, concluded that Australia should act unilaterally to respond to climate change, and that this should involve the introduction of an ETS. The report also contained some preliminary, but detailed, design features for how the ETS should operate. Again, to the surprise of many, on 3 June 2007, prime minister Howard announced that his government would introduce an ETS no later than 2012. Since the Rudd-led Labor Opposition was also committed to the introduction of an ETS, albeit no later than 2011, this ensured that both major parties took an ETS to the 2007 election as formal policy.

Given that prime minister Howard had tasked me with the further development of the ETS, post the Shergold Report, my team and I, still together and operating from the Department of the Prime Minister and Cabinet, threw ourselves into the detailed work, confident that we had bipartisan support for substantive action on climate change mitigation. The issue of adaptation to temperature increases already baked in was, of course, another huge challenge that needed urgent action. The Australian Greenhouse Office and its minister, Malcolm Turnbull, continued their groundbreaking work to implement the National Greenhouse and Energy Reporting Scheme to provide accurate establishment-level measurement of emissions. That office also continued with its leading-edge work on the science of adaptation, but substantive action in this area clearly needed to await the results of the election.

Why had Howard changed his position? Former US vice-president Al Gore's 2006 movie *An Inconvenient Truth* had clearly raised the profile of climate change as a political issue globally and domestically. In Australia, the business community was also becoming increasingly concerned about the action being taken overseas and did not want Australia to be seen to be lagging—they feared they would find themselves penalised, whether through price, access or reputation, in other markets. It is also fair to suggest that, for at least some parts of the business community, there was a degree of concern that, were the 2007 election to result in an ALP victory, such a government might be less open to the needs of business in designing an ETS.

To their credit, though, many in business were conscious of, and shared, the considerable public concern about climate change, triggered by the severe and long-running 'Millennium drought'. Arguably, this led them to 'jump over' the government to a position where they were arguing for the Howard government to drop its longstanding opposition and introduce an ETS well before the prime minister was comfortable in moving in this direction.

The election of the Rudd government in 2007 saw Australia swear in its first-ever climate change minister, Senator Penny Wong, and establish its first Department of Climate Change, to which I was appointed as secretary. Government then immediately embarked on even more detailed analysis of, and consultation on, climate change impacts and mitigation responses. The output of this was reflected in the Garnaut Review,[8] the extensive Treasury modelling exercise,[9] and, ultimately, in the Carbon Pollution Reduction Scheme Green and White Papers,[10] and the eventual CPM legislated as part of the Gillard government's 2011 Clean Energy Future Package. The CPRS and its successor, the CPM, were simply cap and trade emissions trading systems supported by a range of transitional or compensatory mechanisms for business and households.

With the Coalition losing the 2007 election, and the Liberal leadership moving first to Brendan Nelson and then to Malcolm Turnbull, the potential for bipartisanship on climate policy lingered for another two years. But the end of 2009 saw the toppling of Turnbull and his replacement by Tony Abbott, and with it the end of any prospect of bipartisanship around the CPRS in particular and carbon pricing in general.

As Opposition leader, Turnbull had consistently given *in-principle* support to the scheme, subject to seeing the design details, so the Rudd government—via Penny Wong and Greg Combet—had been in negotiations with Turnbull and the then shadow minister for energy and resources, Ian Macfarlane, trying to craft modifications to ensure bipartisan support. On the same December day in 2009 that Turnbull lost the Liberal leadership, the Coalition party room voted to oppose the CPRS legislation that was to be voted on in the Senate the next day.

What happened next was nothing short of a disaster for Australia: The Greens voted with the Abbott Opposition in the Senate to defeat the CPRS.

Some have argued that prime minister Rudd's refusal to engage with Greens leader Bob Brown on the issue played a role. But Rudd, quite rightly, recognised the importance of bipartisanship among the major parties over the issue. Moreover, even had Kevin Rudd been open to engaging with The Greens, the fundamental sticking point would have remained. Specifically, The Greens then—and still today—refused to separate a decision on the design of the mechanism to reduce emissions from the size of the emissions reduction target itself. In 2009, they threw away the opportunity to legislate a mechanism that could be dialled up or dialled down to hit any particular target because they believed the Rudd government's target was inadequate.

It was protest party politics at its worst, highlighting why so few such parties ever move on to seriously influence the direction of economic, social or security policies. In short, considered against Australia's national interest in taking serious action on climate change, it was a classic case of throwing out the baby with the bathwater. But as we shall see, the situation was to get brighter and then worse, much worse.

While it will remain a matter of conjecture, a reputable school of thought holds that if Kevin Rudd had immediately gone to a double-dissolution election after the Senate vote, he would have won the election comprehensively and been able to legislate the CPRS through a joint sitting of the House of Representatives and the Senate. Much like the evolution of its attitude towards Medicare, a number of years of successful operation would likely have muted Coalition demands for the future unwinding of the CPRS. Instead, in April 2010, prime minister Rudd announced the deferral of efforts to introduce the CPRS. The Abbott Opposition, with the help of The Greens, had not only killed the CPRS, they had laid the groundwork for the eventual abolition of the Gillard government's CPM and denied Australia a decade of more effective action on both mitigation and adaptation.

Had The Greens succeeded in having Australia embrace more ambitious abatement targets, such an outcome may have made a modicum of sense. Instead, a decade later, their reward is what they continue to see as a lack of policy urgency, manifestly inadequate targets and no means of efficiently delivering action on climate.

For Australia to have come within one vote of establishing an emissions trading mechanism in 2009 was devastating on many levels. Hundreds of people from Commonwealth and state/territory bureaucracies, business and environmental groups had worked tirelessly, and selflessly, to make the scheme a reality. Compounding the personal sense of wasted opportunity was the fact that, having all but succeeded, my department and I then had to dismantle much of what had been created.

However, history is full of unexpected twists and turns, as 2010 was to illustrate.

Rudd's abandonment of the CPRS led to a sharp drop in public support and he was replaced as prime minister in June 2010 by his deputy, Julia Gillard. The new prime minister did not rule out action on carbon pricing in the future. However, during the run-up to the 2010 election, Gillard gave a public assurance that 'There will be no carbon tax under the government that I lead', while retaining her commitment to the position that the best way to reduce emissions was to price carbon through a market-based mechanism. This subtlety was lost on many!

The result of that election was a minority ALP government relying on the support of The Greens and two key independent MPs. The price of that support, however, was to be a revisiting of the issue of placing a price on emissions.

In September 2010, prime minister Gillard unveiled the Multi-Party Climate Change Committee (MPCCC). Chaired by the prime minister, the committee's members included the new minister for climate change and energy efficiency, Greg Combet; The Greens' Bob Brown and Christine Milne; the Independent Tony Windsor; and a series of advisers including Professor Ross Garnaut, Professor Will Steffen, Patricia Faulkner and Rod Sims. The Coalition was offered two seats at the committee table but, in keeping with his carbon price opposition, Tony Abbott forbade his MPs from participating.

The committee was provided with considerable input from the public service to support its deliberations, particularly from the Department of Climate Change and Energy Efficiency, then led by my successor Blair Comley.[11] The committee ultimately recommended the introduction of a CPM not dissimilar in form to the CPRS, albeit covering a smaller share of the economy.

The CPM was part of a broader Clean Energy Future package negotiated within the MPCCC, which also provided significant tax cuts to households; support for emissions-intensive, trade-exposed industries (EITEs); additional support for a range of other industries (in some cases more than was proposed under the CPRS); a range of measures to support land-based climate action and biodiversity; and measures to improve energy efficiency. The package also led to the establishment of the Clean Energy Finance Corporation and the good work that would follow from that. Additional government support, outside the Clean

Energy Future package, was provided to the steel and coal industries.

Importantly, the CPM was to commence with a carbon price that would be fixed each year for the first three years to allow market participants to become used to the operation of the scheme, before relaxing this element so that the price would float in response to supply and demand for emissions permits, as in the CPRS.[12]

The CPM was indisputably an emissions trading scheme, and on many fronts it was a welcome evolution of the CPRS. But the fixed price element handed the Abbott Opposition the opportunity to claim that it was effectively a carbon tax. This was patently false, but by this stage in the debate facts seemed irrelevant—remember the claims about the $100 leg of lamb![13] The Opposition's claims that prime minister Gillard had misled the public prior to the election were made with brutal effectiveness, aided and abetted by parts of the media, to aggressively undermine her credibility.

With his election win in September 2013, prime minister Abbott's commitment to abolishing the carbon price became a reality. His replacement policy, Direct Action, was simply a reverse auction where the Commonwealth Government paid businesses for abatement. While Direct Action has, of itself, done little direct damage, its very existence was used to argue against the need for other policy responses. Moreover, the amount of abatement that could be purchased was ultimately dependent on the amount of public money that was provided. Ironically, since its supporters opposed business paying for abatement, this abatement was purchased with money raised by taxes on Australian households and businesses—the exact opposite of the 'polluter pays' principle!

As Malcolm Turnbull has noted, Direct Action, adopted after his replacement by Tony Abbott as Opposition leader in 2009, 'was no more than a fig-leaf designed to get the Coalition to the double dissolution election it expected to lose'.[14]

In September 2015, Turnbull replaced Abbott as prime minister. While there was a widespread community expectation that a Turnbull government would take decisive action on climate change, as prime minister, Malcom Turnbull still led a Coalition that had consistently opposed comprehensive action. As a result, Turnbull needed to find sector-specific strategies, rather than a more holistic, whole-of-economy response. Moreover, he needed to focus his actions in areas where emissions reductions were a by-product of delivering other objectives that his Coalition valued. So the Turnbull government focused its efforts on the electricity sector, where the introduction of massive policy uncertainty in 2009 had led to inefficient investment, higher prices for households and business and, in a first for Australia, sovereign risk which decreased our attractiveness as an investment destination.

Following the debate about the security and reliability of the National Energy Market (NEM) after the September 2016 South Australian blackouts, Chief Scientist Alan Finkel was commissioned to undertake an independent review of the NEM. At the time of the release of the terms of reference for the Finkel Review, however, the then environment and energy minister, Josh Frydenberg, indicated that Finkel would also examine the possibility of an emissions intensity scheme—a form of ETS—notwithstanding that this was not in the terms of reference. Not surprisingly, this re-energised the Coalition's climate deniers.

Professor Finkel's report of mid-2017 rejected an intensity scheme in favour of a Clean Energy Target. Having avoided one controversy, the report created a new sense of crisis, since it proposed to provide incentives for generators to produce electricity from a mix of energy sources that had emissions below a given baseline.

With the Renewable Energy Target (RET) seen as a political lightning rod in Coalition politics throughout this time, the CET was divisive within the government, essentially being a RET on steroids. It would have forced a percentage of power generation to come from renewables or other low-emission sources. While this would have delivered lower emissions from the sector, and possibly lower prices over time, it clearly would not have delivered the greater network reliability that was also required. Moreover, it would also likely have hastened the exit of coal from Australia's power system, something that was unacceptable to parts of the Coalition.

Suffice it to say, the CET was dead, notwithstanding the fact that it was not formally put to rest for another year. Its demise came about due to a concerted push by some within the government to treat coal as a form of 'clean energy'. Their strategy was to modify the CET to set the baseline above the emissions associated with a high-efficiency, low-emissions coal-fired power station. No private sector investor would contemplate building assets with a forty-to-fifty-year life that were likely to be abandoned well before then, so there was no private sector support for investment in such plants. As a result, the proponents of this revision to the CET simply suggested the government would fund a plant.

As an aside, the RET has been a key contributor to the challenges faced in developing coherent climate policy. The target provided a significant incentive for investment in both small-scale residential solar installations and large-scale solar and wind.

The RET was hugely popular and effective in kickstarting investment in renewables. The Howard government's initial target when introduced in 2001 was very modest, and this was a world with no commitment to an explicit carbon price. In 2009, the Rudd government increased the target to 41 000 gigawatt hours, or around 20 per cent of the forecast electricity production in 2020, while at the same time introducing an explicit carbon price in the form of the CPRS.

The existence of the RET always carried with it an unanswered policy question: if a government is prepared to put an economically efficient, explicit price on emissions, why is there a need for an additional incentive for renewables? After all, an explicit price sends a signal to both innovators and investors, creating incentives for both technological improvements and the adoption of existing and new low-emissions technologies. It also provides a means to internalise—to incorporate into the price of energy from high-emissions sources—a recognition, in full or in part, of the damage—the externalities—those energy sources create.

The answer is that a RET is ultimately about government picking winners. It subsidises a particular type of energy production, irrespective of whether the market would choose to invest in that technology in either the short or long run. In contrast, placing an explicit price on emissions allows the market to determine which sources of energy to invest in. Introducing a RET on top of an explicit carbon price provides an additional subsidy to renewables. This could make economic sense if it accelerated new technology development, but there is little evidence to suggest that any of Australia's multitude of Commonwealth and state-based RETs have contributed to the rapid fall in global technology costs over the last decade, although they have fostered an Australian industry.

There are really only two rationales. The first is that it provides a vehicle to allow households to contribute to the national abatement task. But that is a political, not economic, rationale. Second, a government may lack the political courage to allow the explicit carbon price to move to a level consistent with its abatement target. In this case, a RET is a way of partly hiding the true cost of achieving a given abatement target.

But back to the climate wars per se. As prime minister, Turnbull had one more roll of the dice.

By the middle of the last decade, the rapid drop in the global price of solar- and wind-generation technologies had made the price relativity between renewables and fossil fuels increasingly unfavourable to the latter, and especially to coal. However, the rapid increase in renewables in Australia's generating fleet had brought with it a deterioration of the stability in the network. Combined with the gold-plating of the transmission network, which had driven up the cost of transmission over the previous decade, the Turnbull government faced a trio of challenges: how to deliver greater reliability while delivering lower emissions and lower prices.

The response was the development of the National Energy Guarantee in late 2017. The NEG immediately garnered strong support from the sector itself, from business more broadly, and from the states. By early 2018, Australia seemed on the verge of at last being able to say it had a sensible, realistic and sustainable policy mechanism that facilitated the market-driven decarbonisation of the energy sector, while also improving reliability and lowering prices for consumers and business.

As we know, this was not to be the case. The climate change deniers in the Coalition rallied in their opposition to the NEG and eventually used its imminent passage as the proximate cause of Turnbull's downfall as prime minister, much as they had done to his leadership of the Opposition over the CPRS in 2009.

LESSONS

With the benefit of hindsight, there seem to be seven lessons to draw from the experience of the last decade. Unless we contemplate these, we will struggle to build support for the needed policy action on climate.

The unavoidable reality is that the Australian economy has developed on the back of our extensive natural resource endowments. This has driven the growth of mining and, within mining, our extensive coal deposits and gas reserves have allowed the creation of major export industries employing large numbers of Australians, especially in regional areas. Moreover, even the composition of our relatively small manufacturing sector has been shaped by access to cheap energy from those fossil fuel resources. Our existing industry structure is, for a developed economy, disproportionately weighted towards EITEs as a direct consequence of our resource endowments.[15] Our population patterns have also been heavily influenced by the geographic distribution of these resource endowments and the industries dependent upon them. It is therefore critical that at the heart of any transition to a lower-emissions future is a robust, long-term bipartisan strategy to smooth the changeover of the affected industries and to safeguard the people of the affected regions. Without a conscious effort to offer a just transition, policy will threaten jobs and regional viability without offering alternatives. There is no faster way to rapidly diminish the chance of broad community support for change.

Additionally, a better strategy is needed to deal with the group of people who claim to be sceptical of the science of climate change, notwithstanding that the basic chemistry has been known for well over a century and a half. These self-described sceptics seem, however, to comprise two quite different groups.

There is one group that passionately rejects the science and nothing will convince them to change their mind—not rational argument, the evidence of more extreme weather events such as bushfires or drought, or emotional pleas. The good news is that, as a share of the community, this group is relatively small. The bad news is that they have disproportionate influence on right-of-centre parties due to their prominence in the media, both mainstream and social media. The challenge for politicians is to recognise that, while the volume coming from a particular megaphone may be high, the numbers shouting into the megaphone are probably low—in short, sometimes it's just best to ignore them and carry on winning the hearts and minds of the rest of the community.

In contrast, true climate sceptics should be open to being convinced. While it is hard to imagine what additional evidence they will find compelling, this position would at least allow a sensible conversation to commence. A true sceptic would embrace the precautionary principle: faced with potentially very long-lived negative consequences, there is value in taking decisions today that maximise the flexibility we have in the future. This was the premise of the 1992 Rio Declaration: when there are threats of irreversible damage, lack of scientific certainty should not be used to postpone cost-effective measures. A true sceptic would demonstrate their scepticism by supporting 'no regrets' actions, thereby taking out insurance against the threat that climate change is real.

There is also a legitimate fear, one that needs to be acknowledged, about moving ahead of the rest of the world, and this has to be addressed in any successful strategy. This fear is usually reflected in two arguments, both of which are entirely legitimate and need to be treated seriously.

Some worry that if we move to reduce emissions, ahead of action by other countries, Australia may end up with a policy apparatus that is out of step with the rest of the world and which may be costly, in terms of jobs and growth, to re-engineer. This concern can best be ameliorated by adopting a comprehensive but adaptable strategy, with mechanisms built in at the outset that allow for eventual linking to what is underway in other countries. The best way to do this is via an explicit CPM that allows for a price link across national borders, say through the creation, sale and use of credible, internationally recognised emissions permits or offsets. This was a key characteristic of the Gillard CPM arrangements.[16]

Others worry that moving ahead of competitors will lead to a loss of international competitiveness, with consequential impacts on jobs, industry and regional viability, and little benefit for the globe. That is, economic activity will be displaced from Australia and replaced by activity elsewhere, with no or negative impact on aggregate global emissions. Both the CPRS and CPM recognised this as a serious issue, and mechanisms to support EITEs were at the heart of each.

In both cases, though, it also needs to be recognised that not acting does not reduce costs. It simply shifts them from mitigation to adaptation and likely increases the total cost in the long run.

If we are to reduce sovereign risk and investor uncertainty, with the incumbent negative impacts on jobs and investment, we need an emissions-reduction mechanism or mechanisms which have bipartisan support. Differentiating between the architecture or mechanism for reducing emissions, and the target those mechanisms may need to hit, is central. If we continue to refuse to separate these two issues, we are destined for another lost decade of action on climate change.

It also needs to be recognised that it is difficult for individual citizens to take a sufficiently long-term view that they are able to appreciate the magnitude of the changes required to the economy. When there is a groundswell of support for action, and for citizens to contribute, that needs to be harnessed. Unfortunately, I think we policy advisers have been too focused in the past on the size of the change, and the technical nature of the responses required, to recognise that at times second-best deviations in policy can help build sustained support for action. It may be, therefore, that policies such as a household-focused RET are valuable in 'oiling the system' to allow other, more effective policies to be adopted.

There is also danger in confusing morality with policy. All policy needs a moral basis, but good policy has to go beyond this to provide a compelling vision of the alternative. The mantra that 'coal is evil' and should immediately be banned, or that we should end the extraction of gas, can provide a warm inner glow, but they do not comprise a policy. The ability to deliver serious emissions reductions while maintaining living standards and satisfying legitimate aspirations for a better life, whether for Australians or others around the world, must be central to policy. Without that, there will be no chance of sustaining a commitment to climate action over the time frames in which we need to operate.

It also needs to be acknowledged that many industrial processes cannot yet, and may never be able to, replace hydrocarbons as a non-energy input. Even if the world achieves net zero emissions by 2050, it is highly likely that we will still be using hydrocarbons in some form. It makes sense, therefore, to concentrate decarbonisation activities in areas where we can have the most rapid impact—reforming the power sector, electrifying industrial processes and personal transport, improving building energy efficiency, increasing the productivity of land through carbon sequestration, and improving agricultural productivity. But this requires a laser-like focus on the most efficient policies in each area, moving away from the mindset implicit in some comments that every sector should contribute equally.

Finally, the problem isn't that we lack a carbon price but that we have too many. Every different state or Commonwealth intervention creates a new, different shadow price on carbon. Every business is using a different implicit price when they assess investments or make decisions. The net result is inefficient and costly. We need a coordinated national effort, not the fragmented, ad-hoc short-termism that passes as policy today.

THE BUSINESS OF RISK

The struggle by the Australian political class to manage the climate issue stands in increasingly sharp contrast to the approach of business leaders and financial regulators globally. Business has concluded they can no longer afford to wait for political leadership. Both here and overseas, investors have placed pressure on firms to set their own strategies and targets to reduce their emissions footprint, while firms also increasingly recognise that failing to do so undermines their own social licence.

From a firm's perspective, they face four main risks due to climate change: mitigation, adaptation, reputation and liability risk.

Mitigation risk arises if governments around the world act so quickly on climate change that they strand a group of assets that are not yet at the end of their physical life. This outcome imposes real value loss on individual firms and investors. The likelihood of this increases the slower that countries are to act on climate change, so that when action does occur, it has to take place over a much shorter period than if planned well ahead of the consequences of climate change becoming indisputable.

Adaptation risk occurs when governments continue to move too slowly and the global average temperature continues to rise, triggering more climate-induced natural disasters that destroy lives and physical assets.

The third risk, reputational, is one that the business community understands well. Companies already deal with activist shareholders on a wide range of issues and investors, and communities more broadly, now increasingly expect action on climate change by firms.

Looming large for business is a further issue: liability risk. The financial consequences for individual firms of being found guilty for the damages caused by their emissions is likely to be significant, although this is more of a potential risk than an actual risk for the time being. It seems only a matter of time, though, before activist groups are successful in 'jurisdiction shopping' for a court that finds in their favour.

It is now clearly recognised that directors of companies need to take account of these various dimensions of climate risk. The work of the Financial Stability Board's Taskforce on Climate-related Financial Disclosure has been instrumental in providing a template for how to report on these issues. It should also be abundantly clear that regulatory attention will only increase, not decrease, over time—as will that of financiers, investors, shareholders, staff and communities.

WHAT NEXT?

On the verge of the 2020s, Australia witnessed first-hand the implications of the continuing failure to develop a resilient climate strategy, one encompassing domestic mitigation and adaptation, and an internationally agreed framework for collective action. The devastating bushfires of late 2019 and early 2020 were remarkable in their ferocity and intensity. While climate change cannot be attributed as the sole cause of any individual extreme event, the failure to address it—to mitigate future temperature rise and adapt to the changes already baked into our climate—will make such events more common over the decades ahead.

Our society—individuals, business, community, politicians—cannot say we haven't been warned. The story of public policy since at least the mid-1990s has been one based on the need to take climate change seriously, to do what we must domestically and internationally to protect our current way of life while meeting the legitimate aspirations of communities, here and abroad, for jobs, incomes and rising living standards. Australia's inability to find a pathway through this morass is significant, not only because climate change is such an enduring challenge, but also because we now face the task of rebuilding our damaged economy after the ravages of the COVID-19–induced recession.

The fall in the pace of economic activity—the change in the growth rate—in 2020 was unprecedented in the ninety years since the Great Depression. The speed of that decline, and the absolute reduction in the level of gross domestic product (GDP), were also without parallel. What remains unknown today is the duration of this drop in economic activity, and the extent of the consequent job losses and social dislocation. Typical recessions lead to a loss of GDP that is recovered over subsequent quarters, but they also result in a rise in unemployment that takes much longer to unwind. Indeed, the early 1990s recession saw Australia's unemployment rate remain above pre-recession levels for over a decade.

Charting a course from the situation in which we currently find ourselves will require innovative approaches to public policy, ways that engage the community over the journey and which build widespread support to stay the course. It will be far from easy and will require all sides of politics to shelve ideological shibboleths in favour of a clear focus on the national interest, for the short, medium and long term. In developing such a pathway, we will have to navigate other issues of considerable complexity, as we were confronting significant pre-existing challenges, both internationally and domestically, prior to COVID-19. These challenges not only make the economic recovery task more difficult, they make it more urgent.

Somewhat counterintuitively, however, these pre-existing economic issues may make the prospects of international action on climate change more difficult while accelerating the prospect of domestic action in many countries. It remains unclear whether Australia will be one of those countries in which action accelerates.

Let me explain what I mean.

THE INTERNATIONAL CONTEXT

If it wasn't apparent before, it is widely recognised now that the world has changed irrevocably over the last decade. Seemingly more immediate issues have either eclipsed climate change or made the achievement of success more difficult.

The period of unchallenged US power that followed the collapse of the Soviet Union is long gone, and the United States and other countries of the West are in their weakest strategic position for decades. The rules-based international order of the postwar era is no longer under threat—it has collapsed!

The United States remains the pre-eminent global power, economically and militarily, but its advantages have been progressively eroded and few other Western nations have stepped forward to help fill the emerging gaps. The Western model of liberal democracy combined with market economics was seriously damaged in the eyes of many by the GFC. The gradual move to more liberal policies, especially in non-democratic regimes, slowed sharply in the face of the policy, regulatory and governance failures revealed by that crisis, which had its epicentre in the United States.

This shift was exacerbated by the apparent robustness and resilience of the Chinese Communist Party model, which absorbed the externally generated economic shocks of the Asian financial crisis and the GFC while leaving the party firmly in control of society and enjoying a still growing economy. Market-based, technology-driven autocracy—techno-autocracy—began to look like a far more appealing model for some countries than liberal democracy.

Economists had long believed there were natural limits to the compatibility of repressive regimes and sustained economic growth. Once catch-up opportunities were exhausted, the costs of repression were expected to undermine the incentive benefits of economic liberalisation. But China's experience meant that it began to look possible that technology had not only increased growth opportunities but also sharply lowered the economic cost of social repression.

Second, while not anticipated at the time, the ascension of Xi Jinping in 2012 changed the direction of policy in China, which became increasingly more militarily muscular, diplomatically assertive and domestically socially repressive. China has also become much more willing to interfere directly in the domestic politics of other countries and is making conscious efforts to draw countries into its orbit using a range of tools that effectively undermine the sovereignty of the affected nations.

China has also shown an increasing willingness to use its economic relationships to coerce countries to its position, with Australia's recent experience bringing it into line with other nations as diverse as Norway, Japan, South Korea, New Zealand, Palau and Canada. Importantly from China's perspective, while the targeted countries are expected to atone for their behaviour, others are expected to see what awaits if they do not comply with China's wishes.

At the same time, there has been far greater questioning of American commitment and willingness to bear the burdens of global leadership, commencing with the Obama administration but accelerating subsequently. The election in 2016 of Donald Trump, with his unique style—a transactional approach to engagement, disdain for alliances, and a cavalier if not imperious attitude towards allies and their interests—sharply eroded one great US advantage over China and Russia in the public's mind: the ubiquity and attractiveness of US soft power. It also led countries to ask: if the US treats allies like that, how can we rely on their commitment to us and our region?

Indeed, as former prime minister Malcolm Turnbull said:

> Trump's deliberate unpredictability generates fear rather than respect, anxiety rather than certainty. America may be stronger in economic and military terms, but its influence is diminished. In fact, under Trump, America seeks less influence, not least by rejecting many of the global institutions created after the Second World War.[17]

Trump's failure to win a second term in 2020 notwithstanding, the reshaping of the post–World War II global power balance—not just the rewriting of the rules but the question of who actually holds the pen—is occurring against the backdrop of strategic competition centred on our region. This is profoundly changing the context in which Australian governments will make decisions around national security, trade and investment, immigration, and even domestic economic and social policy.

Australia is now in open conflict with our major trading partner on a range of issues. These are not minor irritants that will fade but are part of what is, arguably, a tussle over our future sovereignty. There has been a significant escalation in China's approach, which has previously been focused on using the trading relationship to coax Australia into drawing away from our alliance partner, the United States. At the same time, our alliance partner is itself a major cause of concern and source of uncertainty for many countries in the Indo-Pacific, given the inconsistencies in its actions, especially between its trade actions and its strategic objectives.

Successive Australian governments have recognised China's intent, as have others in the region and beyond.[18] In Australia's case, China's actions have led the Morrison government to a more overt statement of intent, illustrated by the calling out of cyber-attacks, a military force posture review, and a new and more significant policy of engaging with the countries of the Pacific. China's overreaching—its lack of transparency during the early stages of the COVID-19 outbreak in Wuhan, its willingness to publicly attack countries that speak out, and the harsh repression of Hong Kong citizens and the rights promised them under the Basic Law—has sent a chilling message. This behaviour places a premium on Australia sustaining close relationships with like-minded countries in our region, particularly Japan, India, Indonesia, Singapore, Vietnam and South Korea.

No country should be so naive as to think of containing China. Its economic rise has been the single most important economic development of the last half-century and is to be welcomed. But working together, Japan, Australia and other such countries are in a better position to balance China, particularly if a more nuanced US commitment to the region were to emerge over time.

It has been clear for many years that the Indo-Pacific, our home region, will be the centre of strategic competition over the decades ahead. This will shape the opportunities and challenges open to Australia, her citizens, businesses and governments. But rather than having the luxury of time to prepare, circumstances have changed far faster than anticipated in recent years. It is no exaggeration to say, as Allan Gyngell has argued, that Australia now faces the most dangerous set of strategic circumstances in modern history.[19]

But what does all this have to do with climate policy?

As Australia has retreated from being a policy leader on climate change, our ability to influence other countries on the issue has diminished. At its heart, climate change is a challenge of the global commons. It is the total stock of greenhouse gases in the atmosphere that determines the climate impacts on individual countries, not what those countries themselves may have emitted. To address this, then, requires concerted global action—collective responsibility and coordinated responses. But achieving such an outcome becomes dramatically harder in a period of low trust between countries, particularly when there is low trust between the world's largest economies and hence its largest emitters.

Add to that the unwillingness of the United States in recent times to exercise its traditional role in being the initiator and coordinator of global action in virtually any sphere—as we have seen so clearly during the COVID-19 medical crisis—and the prospects for rapid action on a further global agreement to tackle climate change look unlikely in the foreseeable future.

While it is not out of the question that the Biden administration might use the threat of instruments such as tariffs to coax other countries back to the negotiating table, or that the clear impacts of climate change on China might lead it to change its strategy, neither of these constitute a viable prospect for delivering rapid international action over the medium term. For Australia, there remains a very real imperative, though, to continue to work with China on technologies to deliver lower emissions. Our export mix is so dependent on China, and so heavily concentrated in resources and EITEs, that our national interest demands we find practical ways to collaborate on climate action no matter the other tensions in our relationship.

But if the prospects for concerted international action are receding, even if only temporarily, the opportunities for domestic action are opening rapidly in many countries.

THE DOMESTIC CONTEXT

Perhaps the defining public policy success story of the last three decades in Australia has been our economic record of twenty-eight years of uninterrupted economic growth prior to COVID-19. The potential of the economy to produce goods and services, and hence to provide jobs and incomes, is determined by three key supply-side factors: population, participation and productivity. While other factors can add or subtract from growth in living standards for lengthy periods, as the terms of trade did during the mining boom of the 2000s, these are generally outside the direct influence of policy and are usually temporary.

For close on a decade, it has been clear that the population and participation drivers are no longer going to be able to provide the support to activity they have provided over the last three decades. Population growth has been expected to slow, notwithstanding immigration; the positive impetus provided by the continuing rise in female participation in the labour market has been diminishing; and male labour force participation has been declining. Indeed, it has long been expected that the first half of the 2020s will see labour utilisation make a flat or negative contribution to growth in living standards. The sharp drop in immigration, both permanent and temporary, associated with COVID-19 will only make this detraction from growth more significant.

With the ending of the terms-of-trade boost, future growth in living standards is entirely dependent on growth in productivity—how clever and efficient we are at combining the factors of production, land, labour, capital and technology. Moreover, it has been apparent that if Australians want to maintain the growth rate in living standards experienced over the last three decades, actual productivity would need to grow at nearly twice its long-run growth rate over the coming decade. Unfortunately, labour productivity growth in Australia has slowed sharply over the last twenty years, and in recent years has averaged only around one-third to one-half of its long-run growth rate.[20]

The reasons for this slowdown are complex. Some have attributed it to measurement problems, but this seems an implausible explanation for the extent of the slowdown. Across the world, productivity growth has slowed, suggesting the pace at which the productivity frontier is expanding may be slowing, notwithstanding technological breakthroughs across a range of areas. The Organisation for Economic Co-operation and Development has also suggested that the gap between the firms that constitute the leading and trailing edges of productivity within individual sectors has also widened. The interesting question is what has contributed to this slowing of the rate of technological diffusion.

In Australia, it is also partly a consequence of where jobs growth has been occurring, in what can be categorised as 'the caring sectors'. These are characterised by low productivity levels and low productivity growth rates, hence lower wage rates and slower wage growth. Indeed, the slowing in aggregate labour productivity explains about half the slowing of wage growth since the end of the mining investment boom in 2012.[21]

Whatever weight history gives to these myriad reasons, many in the community have seen the sharp slowing of wage growth over the last decade as evidence of someone else gaining at their expense. This has further exacerbated the erosion of trust in our social compact.

An ageing population brings additional challenges. As the median voter ages, they are likely to become more resistant to the changes involved in major, productivity-enhancing reforms. Moreover, as the share of the population in older age groups grows, those age groups are likely to carry more political influence, meaning they are likely to seek, and receive, disproportionately more support from government than other cohorts in society.

It is not surprising, therefore, that it has become harder to build political consensus for the sorts of policies needed to drive faster growth in living standards. Such policies have been seen by some as risking political capital, something they have been hesitant to spend in a low-trust environment.

These impacts can be ameliorated at the margin, though not completely overcome, through high-quality immigration. But, as we have seen, that throws up geographically concentrated population pressures that bring other demands for government action. They can also be eased through a more forthright embrace of technological change, but the experience of recent years suggests that merely mentioning innovation can create deep concern in parts of our community about the future of work.

So, even before COVID-19, Australia faced significant economic and political challenges simply to sustain what we had become used to. The pandemic has made those challenges harder and the imperative to address them more urgent.

The question now in front of governments, both Commonwealth and state/territory, is how to jump-start the Australian economy. Clearly, the traditional policy levers, such as tax reform and infrastructure spending, have a key role to play, but they are unlikely to be sufficient. The question is what else may assist.

The International Monetary Fund (IMF) has called for governments to invest selectively in sectors and businesses that will benefit from digitisation and green climate policies. Poorly designed policy over the last decade has created sovereign risk in the Australian energy sector, so proceeding down this route requires care. In particular, all pathways to a low-emissions environment require the deployment of large amounts of risk capital over many decades in both the development of new technologies and their uptake. Any government intervention needs to be carefully designed to not cut across incentives for the private sector.

That said, Australia needs a more activist approach to innovation policy more broadly, and to climate-related issues in particular. Exciting possibilities exist to turn some of our existing fossil-fuel-heavy industrial users into lower-emission industries through the electrification of industrial processes. The electrification of personal transport can be accelerated. Energy-efficiency opportunities abound. We have a great chance to get in on the ground floor of the hydrogen economy. And if we do all of these, we can rapidly decarbonise our energy system, particularly the electricity sector, and drive down costs for households and business.

In contrast, recovery packages that look backward, that fail to grasp these prospects, not only miss the opportunity to create jobs and growth with co-benefits like energy security and reliability, they will lock in escalating risks for decades to come.

In terms of concrete actions, Australian governments have an abundance of guidance, both domestically and internationally. The International Energy Agency (IEA) head, Dr Fatih Birol, has been explicit: 'Governments have a once-in-a-lifetime opportunity to reboot their economies and bring a wave of new employment opportunities while accelerating the shift to a more resilient and cleaner energy future'. The IEA has provided governments with a smorgasbord of options

> designed to: (1) accelerate the deployment of low-carbon electricity sources like new wind and solar, and the expansion and modernisation

> of electricity grids; (2) increase the spread of cleaner transport such as more efficient and electric vehicles, and high-speed rail; (3) improve the energy efficiency of buildings and appliances; (4) enhance the efficiency of equipment used in industries such as manufacturing, food and textiles; (5) make the production and use of fuels more sustainable; and (6) boost innovation in crucial technology areas including hydrogen, batteries, carbon capture utilisation and storage, and small modular nuclear reactors.[22]

As is already apparent from the stimulus packages in many European member countries, some are embracing this challenge and addressing the need to rebuild their economies while tackling the climate challenge, placing themselves in a better position for a post-COVID-19 economy.

LOOKING FORWARD, LOOKING BACK

The United Nations describes the central aim of the Paris Agreement on climate action as being 'to strengthen the global response to the threat of climate change by keeping a global temperature rise this century well below 2 degrees Celsius above pre-industrial levels and to pursue efforts to limit the temperature increase even further to 1.5 degrees Celsius'.[23] Realising this goal requires a significant reduction in emissions this decade and the achievement of net zero emissions by 2050. The IEA suggests, however, that even if governments implemented all their existing Paris commitments, global average temperatures are likely to rise by between 2.7 and 3.5 degrees Celsius. In short, governments around the world need to both increase their level of ambition and then put in place policies consistent with that ambition.

It is not easy to be optimistic in this regard. Of vital importance is an emissions budget, for it is the stock of greenhouse gases in the atmosphere, not how much is added in any one year, that is critical. For this reason, it is not plausible to commit to doing a lot at a later date in order to avoid action today. Unfortunately, countries have set their abatement targets in terms of percentage reductions at a point in time relative to a base year; for example, Australia's commitment is to reduce emissions in 2030 by 26–28 per cent compared with 2005 levels.

Ultimately, whether the emissions come from Melbourne or Moscow, the impact is the same. While it might be true on one level that Australia can do only so much because it represents 1.3 per cent of total emissions, it is also among the world's highest emitters on a per capita basis. As a nation, we have the capacity to deploy moral suasion on others to act, but we also have a duty to be a good global citizen and act ourselves.

Unfortunately, our attitude to the use of carryover credits to meet Paris Agreement targets undermines our position globally. What is little understood is that the use of carryover credits makes Australia's future abatement task harder, and more costly.

Carryover credits have arisen because Australia overachieved relative to the initial Kyoto targets. Because of the focus on point-of-time targets, if Australia uses carryover credits to achieve its

2030 target, its actual level of emissions that year will be higher than the target level. If the world then embraces greater ambition—say net zero emissions by 2050—Australia will be starting much further away and hence will require more net abatement, at greater cost, than if it had met the 2030 target without recourse to carryover credits. This use of point-in-time estimates highlights two peculiarities about Australia's climate debate: first, the business as usual (BAU) pathway relative to other developed economies; and second, the sensitivity of attitudes towards the ambition of targets to the language in which they are framed.

Australia's BAU—the baseline scenario that assumes no mitigation policies or measures will be implemented beyond those already in place—is extremely unusual compared with other developed economies. Australia typically grows faster than almost any other developed economy—much faster than Europe or Japan, and generally a little faster than the United States and Canada. Importantly, Australia's population has also grown faster than the rest of the developed world. So, even if emissions per capita decline significantly, a rising population means emissions in total don't fall commensurately.

This can be illustrated with reference to the recent targets. Recall that The Greens rejected the CPRS on the grounds that the Rudd government's target—to reduce emissions by 5 per cent below the 2000 levels by 2020—was inadequate. At the time of the commitment, however, it would have required Australia to reduce emissions per capita by one-third between 2000 and 2020. Expressed that way, it would have seemed very ambitious. Similarly, at the time of its announcement, the Abbott government's 2030 target—a reduction of 26–28 per cent below 2005 levels by 2030—was the equivalent of Australia reducing its emissions per unit of GDP by almost two-thirds, or its emissions per capita by a half.

None of that is to argue that more ambition is unnecessary. But it highlights the importance of carefully framing, and describing, the ambition of policy reforms.

THE PATH FROM COVID-19

It was noted at the outset of this book that the success of the 1980s and 1990s reform agenda reflected three key factors:

- a sense the status quo was unsustainable
- a consensus in favour of trying new approaches
- a political culture prioritising the national interest.

The discussion of Australia's experience has highlighted that none of these three criteria have been satisfied in the context of the ongoing debate on climate change.

The need to chart a growth path from COVID-19 opens the window of opportunity to accelerate the decarbonisation of the economy, to foster new industries and new jobs—in short, to simultaneously address two great challenges, rather than seeing them in conflict.

There are two pathways by which this might occur: governments may lead in this direction, or business and the broader community may strike out, leaving politicians to play catch-up, much as happened to the Howard government in the mid-2000s. But even if one of these two scenarios comes to reality, the world is still likely to lack a coherent response to the climate challenge, with the new international context making it dramatically harder to achieve consensus on collaborative, coordinated action. The other risk is that governments may choose to turn their backs on these opportunities, swamped by the challenge of economic recovery and lacking the imagination, or too overcome by timidity, to chart a new course.

While the jury currently remains out, there was little to suggest pre-COVID-19 that Paul Kelly's criteria might come together, whether around climate or broader economic policy reform. It would be an extreme irony if a medical emergency created the circumstances for a renewed vigour in tackling both these challenges, but one can hope.

Certainly, the decision by the United States to withdraw from its traditional multilateral or plurilateral leadership role has highlighted the structural weakness of the international architecture. COVID-19 has shown that US-led coordination is indispensable for action, even if the United States cannot lead in the unchallenged way it has for the past thirty to forty years. A globally weakened United States, and Europe, will see the strategic competition playing field tilted in China's favour. It would be odd were China not to use the opportunities this creates. The era of heightened strategic competition will be something Australians will now have to accept as part of the 'new normal'.

Returning to climate change, this highlights another key challenge. Addressing any of the challenges to the global commons—climate change, pandemics, mass irregular migration, ocean pollution—requires concerted global action. In a world marked by an absence of leadership, low trust between nations and clear strategic competition between the two major players, it seems the likelihood for an effective coordinated response to these problems is diminished, not enhanced, despite the threats they pose.

COMPARING CRISES

In thinking about how crises unfold, it can be useful to break them into three phases and to consider the issues and lags associated with each. In simple terms, these three phases can be understood as:

- recognition lag—the time between an event becoming apparent and policymakers recognising they need to change course
- reaction or gestation lag—the time taken to work out what that course correction should be and then to implement those changes
- response lag—the time between the implementation of new policies and those policies taking effect.

During the 1991–92 recession, each of these lags turned out to be relatively lengthy. Government was slow to acknowledge that the economy was going into a marked slump, notwithstanding partial data showing the economy slowing during

1990. The recognition lag was particularly lengthy due to the bureaucratic and political commitment to what turned out to be a major contributor to the recession: the belief that high interest rates could simultaneously reduce excess demand (which they can), curb speculative behaviour in the property market (more difficult) and help reduce Australia's current account deficit (even less likely).

This policy decision was exacerbated by the Reserve Bank's (RBA) need to gain the agreement of the treasurer to changes in interest rate settings, and the government's reluctance to speak openly about what was being done. Monetary policy has two channels by which it influences the real economy: an instantaneous announcement (or confidence) effect, and an eventual impact on liquidity, which typically comes with 'long and variable lags'. With a muted announcement, market participants, businesses and individuals found it difficult to discern the direction of policy. This, in turn, resulted in policy being tightened more than needed in response to an economy apparently stubbornly failing to respond to policy changes.

It is worth noting that the world's central banks now commonly give clear guidance to market participants about the direction of policy, to the extent that actions to implement policy—for example, stepping into the market to purchase bonds to drive down interest rates—are sometimes not actually required to achieve the announced objective. The credibility of the central banks induces market participants to act *as if* the actions had already occurred, in the full knowledge they *will* occur if the interest rates do not move to the target.

Returning to the recession, when it was finally accepted that action was needed, the effort required was underestimated and the initial response in November 1991 was clearly inadequate. It took until February 1992—after Paul Keating had replaced Bob Hawke as prime minister—to release a comprehensive response strategy, titled *One Nation*. Unfortunately, unemployment was already rising steadily, eventually reaching 11.2 per cent in the December quarter of 1992. Moreover, many of those who were out of work, especially males over the age of fifty-five, never found new jobs. For a variety of reasons, the combination of policies in *One Nation* was such that the bulk of the stimulus did not hit the economy until late 1993, and was then so stimulatory that by mid-1994 monetary policy needed to be tightened sharply to slow the pace of activity. Accordingly, the recognition lag was around twelve months, the reaction lag was a bit over six months and the response lag was around eighteen months.

In contrast, while regulators globally missed the build-up of imbalances and regulatory failures that contributed to the GFC, the uncertainties and fragilities that were emerging in financial markets were the topic of conversation among policymakers in Australia and overseas from early 2008. When Lehman Brothers collapsed on 15 September that year, the cascading impact on markets and other market players was significant. The Australian authorities responded by banning short-selling, buying residential mortgage-backed securities, cutting interest rates, guaranteeing bank deposits, and delivering a fiscal stimulus of close to 1 per cent of GDP—all within one month of Lehman's collapse.

It is interesting to consider why the response was so different to that of the earlier recession.

In Australia, the RBA had become 'instrument independent' after the early 1990s recession, meaning that while the goal of monetary policy (the 'target' of keeping inflation to 2–3 per cent on average over the cycle) was agreed between the government and the Reserve Bank, and formalised in a written agreement between the two, the RBA had freedom regarding how to deliver that goal. Alongside this, the leadership of Treasury under its then secretary, Dr Ken Henry, had considered in-depth the evolution and consequences of the previous recession and was committed to being as well placed as possible to respond if circumstances turned negative. Recognising that only he and I among the leadership team had been in influential positions in the early 1990s, in 2004 Henry commissioned from staff a set of scenario analyses to assist the department in planning for how it might best advise government to respond to a future downturn. While this resulted in some questioning in the Senate as to whether Treasury was expecting such an event—which it was not—it was accepted that this was not only good training for staff but prudent risk management behaviour some twelve years after the last recession.[24]

Just how prudent that was became apparent during the GFC, where that scenario work, and the extensive leadership discussions it had provoked within Treasury, allowed Ken Henry to advise prime minister Rudd to 'go early, go hard, go households'.

The nature and magnitude of the GFC was also notable in bringing forth an unprecedented degree of policy collaboration and coordination, including the establishment in 2008 of the G20 at the leaders level.[25] During 2008, 2009 and 2010, those leaders met twice-yearly and spoke regularly in-between, while policy officials and regulators embarked on unparalleled levels of engagement.

The experience of the early 1990s recession had made a significant impact on the thinking of both Ken Henry and myself. In particular, we both understood that the inevitable rise in unemployment in a recession lags the change in GDP growth, so that in attempting to respond to a downturn with activist fiscal policy, it is critical to pursue actions that take effect with minimal lags. In contrast, key components of *One Nation* took considerable time to be put in place—this was especially true of some of the infrastructure initiatives.

If we consider Australia today in the context of the medical crisis of COVID-19, the recognition, response and reaction lags have been very short and, broadly, highly successful. The economic recognition and reaction lags have also been remarkably short—only a matter of weeks from the start of health-driven closures of part of the economy to the release of both the first and second economic packages designed to support the community through the medical component of the crisis. Again, they seem to have been highly successful, both in sustaining household incomes during the crisis and in keeping workers attached to their employers.

While it is inevitable that unemployment will remain at elevated levels post the medical emergency, the initiative to establish JobKeeper has been important in minimising the likely unemployment consequences and, in particular, the sidelining from future work of an entire cohort of Australians, as occurred in the 1990s. Unfortunately, history is likely to show that it was far easier to establish JobKeeper than to exit from it. How that exit is managed will have serious ramifications for the level and duration of unemployment.

The next step, economic recovery, requires measures to deliberately stimulate economic activity. The most recent budget emphasised traditional infrastructure projects, with a focus on being 'shovel ready'—that is, where the lag to beginning work is short—and some areas of renewed reform focus. What is striking about the current circumstances is the emphasis on stimulating activity in traditionally male-dominated industries. The impact of this slowdown has, however, been highly concentrated in sectors where women and young people are over-represented. The long-term consequences of the crisis are, therefore, likely to be substantial in terms of derailing the closing of the gender pay gap and the gap in labour force participation rates, in terms of the distribution of superannuation savings for retirement, and in the lifetime incomes of young, recent entrants to the labour market.

While the next budget may introduce further measures, taken together, these responses will need to withstand scrutiny on two issues: are they sufficient to encourage growth and jobs in the short to medium term, and are they in areas that build Australia's capacity to grow and compete in a more competitive global economy over the decades ahead?

To achieve the latter requires going beyond traditional agendas to also focus on encouraging innovation, a more rapid move to the digitisation of the economy, and identifying opportunities to decarbonise by encouraging the market-driven energy transition that is already underway, while also encouraging new low- or zero-carbon industries. It will also require a clear-eyed strategy on how to position Australia economically and strategically in the Indo-Pacific, a strategy not driven by US–China strategic competition but by a recognition that our future is intertwined with that of the other countries of the region.

The policy response provides the opportunity to both address the need to reorient the Australian economy towards a decarbonised future and to stimulate new jobs, industries and export opportunities. The IEA, supported by IMF modelling, has illustrated how this could occur.[26] It is clear that many countries in Europe are embarking on such a 'green recovery'. The newly elected Biden administration is likely to embrace elements of the so-called 'Green New Deal'. If Australia fails to embrace at least some elements of these strategies, it risks being left behind as the new industries of a low-carbon world emerge.

Whatever ultimately comprises the policy response, the great unknown will remain the speed of the third lag—the time it takes the economy to respond to these measures. That will also be influenced by the speed of the global economic recovery and its feedback to the Australian economy through trade, investment and confidence. The IMF has warned of deep scarring for the global economy, with the downturn in some key developed economies likely to be deeper than in Australia, given the success to date in managing both the medical and economic response.

SUMMING UP

Considering the nature of the crisis, what has been striking about the COVID-19 pandemic has been the absence of any effective, globally coordinated medical or economic response. We have witnessed a medical emergency impacting multiple countries simultaneously, recessions in many countries caused by the response to the emergency, and a transmittal of slowing growth, even to countries minimally affected by the virus itself, through weaker trade and investment. At any other time in the postwar era, international responses would have been prepared to deal with the medical emergency—witness the international response to Ebola, severe acute respiratory syndrome (SARS) and Middle East respiratory syndrome (MERS)—while international bodies and groupings would have been calling meetings of world leaders to collaborate on and even coordinate economic responses. The absence of coordinated action is again reflective of a lack of trust between countries and a conscious US decision to withdraw from the international order it crafted in the post–World War II era.

The period ahead will be a time of great economic, social and security uncertainty. Let us hope that we can navigate this successfully and emerge with a stronger, more economically vibrant and socially cohesive Australia, continuing to integrate with the countries of our region to create a safe, secure and prosperous Indo-Pacific.

But this will not occur simply because we wish it to be so. It will require hard work, hard decisions and a clear focus on Australia's national interest.

ACKNOWLEDGEMENTS

This book draws in part from a video conversation with journalist and author Marian Wilkinson for the University of Technology Sydney ('Does Climate Policy Have a Future in Australia?', 19 November 2019) and a podcast interview with *The Guardian*'s Katharine Murphy ('Australian Politics Live', 19 December 2019). I would like to thank Marian, Katharine, UTS and *The Guardian* for their willingness to allow me to draw on that material. This work has also benefited from valuable comments from Allan Gyngell, Ken Henry, Heather Smith and Marian Wilkinson. All errors are my own.

NOTES

1 Paul Kelly, *Triumph and Demise*, Melbourne University Press, Melbourne, 2014, p. 70.

2 Martin Parkinson, '*The Lucky Country: Has It Run out of Luck?*', Working Paper 247, The Griswold Center for Economic Policy Studies, Princeton University, Princeton, NJ, September 2015.

3 Although Australia had implemented an individual transferable quota system for Southern bluefin tuna.

4 Notwithstanding initial support for the Kyoto Protocol, the Howard government announced in June 2002 that it did not intend to ratify the agreement. It was not until the election of the Rudd government in 2007 that Australia provided ratification. This occurred on 3 December that year and was symbolically important as the Rudd government's first policy action.

5 The four papers were titled 'Establishing the Boundaries', 'Issuing the Permits', 'Crediting the Carbon' and 'Designing the Market'.

6 The paper was titled 'Pathways and Policies for the Development of a National Emissions Trading System for Australia'.

7 Seven of the twelve members were from the business community. The remainder were secretaries of Commonwealth departments.

8 Ross Garnaut, *The Garnaut Climate Change Review*, Cambridge University Press, Melbourne, 2008.

9 Australian Government, *Australia's Low Pollution Future: The Economics of Climate Change Mitigation*, Commonwealth of Australia, Canberra, 2008.

10 'Carbon Pollution Reduction Scheme Green Paper' (July 2008) and 'Carbon Pollution Reduction Scheme White Paper: Australia's Low Pollution Future' (December 2008).

11 I was, by this time, secretary to the Treasury.

12 Australian Government, *Securing a Clean Energy Future: The Australian Government's Climate Change Plan*, Commonwealth of Australia, Canberra, 2011.

13 Then National Party Senate leader Barnaby Joyce claimed that the price of a leg of lamb, then less than $20, would rise to $100 with the introduction of a price on carbon.

14 Malcolm Turnbull, *A Bigger Picture*, Hardie Grant, Melbourne, 2020, p. 168.

15 It is somewhat ironic that the policy reforms of the 1980s and 1990s, which were intended to reinforce this pattern of specialisation, were not followed with subsequent phases of reform that could have built new forms of endowment and delivered a more digitised and diversified economic model for Australia. It was, and still is, easier to just do more of the same rather than make the investments needed to deliver a modern, more diversified economy.

16 For this to be effective, though, there needs to be robust confidence in the integrity of the permits—that they actually deliver what they claim to represent. In the case of the CPM, there were concerns that the use of international permits would see the short-term fixed price exceed the long-term internationally influenced price due to insufficient safeguards on the international permits.

17 Isabella Kwai, 'Letter 158: Beer with Bella—Malcom Turnbull', The Australia Letter, *The New York Times*, 22 May 2020.

18 See, for example, prime minister Turnbull's speech to the sixteenth International Institute for Strategic Studies (IISS) Asia

Security Summit in 2017, and Singaporean Prime Minister Lee Hsien Loong's address to the same summit in 2018.

19 Allan Gyngell, 'Saving the World Order: Picking up the Pieces', *The Economist*, 4 August 2018; 'The Company We Keep: Risk and Reward in the Time of Trump', *Australian Foreign Affairs*, vol. 1, October 2017, p. 54.

20 Productivity Commission, *PC Productivity Insights: Recent Productivity Trends,* Commonwealth of Australia, Canberra, February 2020.

21 Ibid., p. 25.

22 International Energy Agency, *Sustainable Recovery: World Energy Outlook Special Report*, IEA, Paris, June 2020, https://www.iea.org/reports/sustainable-recovery (viewed October 2020).

23 United Nations Framework Convention on Climate Change, 'What Is the Paris Agreement?', 2020, https://unfccc.int/process-and-meetings/the-paris-agreement/what-is-the-paris-agreement (viewed October 2020).

24 Coincidentally, it was not long after this exercise that the present Secretary to the Treasury, Dr Steven Kennedy, and colleagues Jim Thomson and Petar Vujanovic released 'A Primer on the Macroeconomic Effects of an Influenza Pandemic' (Treasury Working Paper 2006-01, February 2006).

25 Prior to this, the G20 had been a forum solely for finance ministers (that is, treasurers) and central bank governors since its establishment in 1999.

26 The IEA's *Sustainable Recovery Plan* (Paris, June 2020) combines economic recovery with building cleaner and more secure energy systems, fostering energy efficiency, and improving industrial and transportation systems.

IN THE NATIONAL INTEREST

Other books on the issues that matter:

Bill Bowtell *Unmasked: The Politics of Pandemics*

Rachel Doyle *Power & Consent*

Kevin Rudd *The Case for Courage*

Don Russell *Leadership*

Scott Ryan *Challenging Politics*

Simon Wilkie *The Digital Revolution: A Survival Guide*